The Royal Botanic Gardens
KEW

SOUVENIR GUIDE

LONDON HER MAJESTY'S STATIONERY OFFICE

Note: This was first issued as a publication of
the Royal Botanic Gardens, Kew

Designed by HMSO Graphic Design

Printed in the United Kingdom for
Her Majesty's Stationery Office
by Shenval Print Ltd, Harlow

Dd 239585   C1000   5/87

Contents

The reference numbers in the text are
grid references to the Map

General Information

The Gardens are open daily (except Christmas Day, and New Year's Day) from 9.30am. The closing hour varies from 4pm–6.30pm on weekdays and 4pm–8pm on Sundays and Public Holidays, depending on the time of sunset.

The Museums open at 9.30am and the glasshouses at 10am; their closing hours also vary, but they are never open later than 4.30pm on weekdays and 5.30 pm on Sundays. The Museums are closed on New Year's Day, Good Friday, Christmas Eve, Christmas Day and Boxing Day.

Visitors are asked not to touch the plants and trees. The climbing of trees is not allowed. The playing of radios, cassette players and musical instruments is not permitted. Ball games and other games and sports activities are not allowed in the Gardens. Guide dogs only are permitted.

Charges for admission

Visitors 50p (no reduction for parties). Children under 10 years free. Wheelchairs 50p (ticket admits occupant and attendant).

Season tickets are available for 12 months and may be obtained at the Gates or on application to Public Relations at a charge of £5. They are not transferable and may be revoked at the discretion of the Trustees of the Royal Botanic Gardens, Kew.

School parties (including a teacher or teachers) are admitted free (except on Saturdays, Sundays and Public Holidays) by voucher which can be obtained by written application to the Information and Exhibitions Division, Royal Botanic Gardens, Kew, Richmond, Surrey, or telephone: 01 940 1171 ex. 4615, giving not less than seven days' notice. Teachers are asked to ensure that school parties are adequately supervised at all times.

Guide lecturer service

Guide lecturers are available to take parties (not exceeding 20 people per guide) around the Gardens, glasshouses and Museums on Mondays to Thursdays only. It is essential that application for these tours is made in advance by telephoning the Information and Exhibitions Division: 01 940 1171 ex. 4615. Early booking is necessary to avoid disappointment.

Wheelchairs

A limited number of wheelchairs are available for hire at a charge of 50p. These can be booked in advance by writing or telephoning the Sergeant's Office. They may be taken into the glasshouses and Museums where access is available, but should not be taken onto the lawns. Electrically

◀ **The Broad Walk showing rhododendrons and spring bedding**

driven wheelchairs only of the kind permitted on the footways under the Chronically Sick & Disabled Persons Act 1970 are admitted to the Gardens.

🔔 Photography and painting in the glasshouses
Visitors wishing to use tripods or easels in the glasshouses (permitted on weekdays only), must apply in advance for a permit from Public Relations.

Permission to film, photograph or paint for commercial purposes in the Gardens, glasshouses and Museums is at the discretion of the Director, and must be obtained in advance through Public Relations.

Visitors are particularly asked not to handle the plants.

🔔 Refreshments
These may be obtained either at the Refreshment Pavilion (K6) during the summer season (Easter to the end of October) or at the Refreshment Kiosk (Tea Bar D14) which is open throughout the year from 10am to one hour before the Gardens close. Applications for special terms for parties, etc., should be made *direct* to the Manager, Refreshment Pavilion, Kew Road, Kew, Richmond, Surrey. Telephone: 01 940 7177. There are several hotels, public houses and restaurants in the vicinity.

🔔 Toilets: ‡ with facilities for the disabled
*closed in winter

Ladies:
1. ‡Near Orangery E16
2. ‡East of Waterlily Pond E5
3. ‡East of Refreshment Pavilion L6
4. ‡East of Palm House Pond K13
5. Near Brentford Ferry Gate D13
6. ‡Refreshment Pavilion *K6

Gentlemen:
1. ‡Behind Orangery E16
2. ‡East of Waterlily Pond E5
3. ‡East of Refreshment Pavilion L6
4. ‡East of Palm House Pond K14
5. Near Brentford Ferry Gate D13
6. ‡Refreshment Pavilion *K6
7. ‡Near Victoria Gate K11

🔔 Parking
Cars may park in the vicinity of Kew Green or in Queen Elizabeth's Lawn car park, near Brentford Gate, which is reached through Ferry Lane (off the north side of Kew Green). Coaches should drop their passengers at the Main Gate.

By bus. Numbers 27 and 65 (on summer Sundays NO. 7).
By underground. District Line to Kew Gardens Station.
By North London line. Richmond train to Kew Gardens Station.
By BR Southern Region to Kew Bridge Station.
By riverboat. From Westminster Pier to Kew Pier, near Kew Bridge and
the Main Gate (April to October only).

History and functions

The Royal Botanic Gardens consist mainly of two estates which once
belonged to the royal family. In the 1720's George II and Queen Caroline
lived at Ormonde Lodge (or Richmond Lodge) on the Richmond estate,
which ran by the river between Richmond and Kew. In the early 1730's
their son Frederick, Prince of Wales, took a lease of the neighbouring Kew
estate, which extended from Kew Green to the present southern boundary
of the Gardens near the Pagoda. Prince Frederick died in 1751 and in
1759 his widow, Augusta, Dowager Princess of Wales, mother of King
George III, started a botanic garden of about 3½ hectares (9 acres) on the
land south of the Orangery. William Aiton was Princess Augusta's head
gardener and Lord Bute her botanical adviser. Sir William Chambers
designed a number of buildings for the estate, including the Orangery,
Pagoda and Ruined Arch.

George III inherited the Richmond estate on his grandfather's death in
1760 and took over the Kew estate on his mother's death in 1772; the two
properties were eventually joined together. Sir Joseph Banks became the
unofficial director and it was largely due to him that the Botanic Garden
became so famous. It was at his instigation that young men were sent out
to many parts of the world to collect plants for Kew.

In 1841, after a period of decline, the Botanic Garden was handed over
to the state as the result of a Royal Commission; soon afterwards the royal
family gave further large areas of the estate, which increased the size of the
Gardens to over 81 hectares (200 acres). The first director was Sir
William Hooker, who started both the Department of Economic Botany
and Museums and the Herbarium and Library. The Jodrell Laboratory
was founded in 1876, when Sir Joseph Hooker, son of Sir William, was
Director. Queen Charlotte's Cottage and its grounds were given by Queen
Victoria in 1898 to commemorate her Diamond Jubilee in 1897.
Cambridge Cottage (part of which is now the Wood Museum) and its
garden were presented by Edward VII on the death of the last Duke of
Cambridge in 1904. The Gardens today are slightly more than 121

Augusta, Princess of Wales

hectares (300 acres) in extent and lie on the flood plain terrace of the River Thames. The soil consists of river gravels and sands at least 9 m (30 ft) deep, overlying London clay.

In 1965 the Royal Botanic Gardens took over Wakehurst Place, an estate of 187 hectares (462 acres) in Sussex, which was bequeathed to the National Trust on the death of the owner, Sir Henry Price, in 1963. The gardens are leased from the Trust at a peppercorn rent and contain many plants from S. America and Australasia which are rarely seen out-of-doors in England.

The Gardens were formerly part of the Ministry of Agriculture, Fisheries and Food, but since April 1984 they have been administered by an independent board of trustees and funded by direct grant from MAFF Contrary to popular belief, the Gardens' primary purpose is not to serve as a public park, but as a scientific institution. An important function is the accurate identification of plants from world-wide sources. The Gardens also act as a centre for the distribution of economic and decorative plant material.

Kew is increasingly involved in conservation. With thousands of species of plants all over the world threatened with extinction, some of those in danger are carefully grown in the Gardens or their seeds are stored in the Seed Bank situated at Wakehurst.

🌣 The School of Horticulture

At present 16 students are selected each year for the Horticultural Course leading to the Kew Diploma. The course extends over three years and includes practical work in the Gardens as well as instruction in the laboratory, drawing office and lecture theatre. The Diploma is a professional and technical qualification of high standing. Entrants, selected competitively, average 22 years of age on entry.

Further details of the course may be obtained by writing to the Supervisor of Studies, Royal Botanic Gardens, Kew.

In addition to the Living Collections and Administration there are three other Divisions:

🌣 The Herbarium and Library

The Herbarium is housed in the large block of buildings to the right of the Main Gate. The collection of dried and pressed plants, about 5 million specimens, is one of the largest in the world. It is not open to the public, but *bona fide* students and research workers may study the collections on application. Housed in the same building is the Library which ranks as one of the best and richest botanical libraries in the world with over 120,000 volumes and extensive collections of archives, illustrations, maps, etc. It is available only to *bona fide* students, on application.

🌣 The Information and Exhibitions Division

The Division is responsible for all information and educational services. It also presents exhibitions and educational displays in the Gardens and elsewhere. From April 1987, the Division will be housed in the new Sir Joseph Banks Building which contains offices, storage for the Economic Plants Collection and exhibition space.

🌣 The Jodrell Laboratory

This Division undertakes research on the anatomy, cytology, biochemistry and physiology of plants, the last section being at Wakehurst Place.

Further information on the work of these Divisions may be found in the orientation display area in the Orangery.

Buildings and landmarks

♨ Pagoda K3

The Pagoda is probably Kew's best known building and was designed for Princess Augusta (see historical section) by Sir William Chambers in 1761–2. Chambers was the architect of Somerset House and wrote a famous essay on oriental gardening. The building is 49.69 m (163 ft) high and octagonal in shape, has ten storeys and took only 6 months to complete. It is situated in the south-east corner of the Gardens, near Lion Gate. For reasons of safety it is not open to visitors.

♨ King William's Temple H9

This temple is between the Palm House and Temperate House and was built at the end of King William IV's reign in 1837. A series of plaques around the interior commemorates the battles fought by the British Army between 1760 and 1815. Many visitors find it a convenient place to have their picnics on wet days.

♨ Temple of Aeolus J14

This temple, situated on the artificial mound by the Palm House Pond and north of Museum I, was designed by Sir William Chambers and built between 1760–3. It was rebuilt in 1845 by Decimus Burton, as the original building had fallen into a ruinous state.

♨ Temple of Bellona K11

Set on a slight rise to the south of Victoria Gate, this temple was designed by Sir William Chambers and built in 1760. In spring the ground in front is thickly covered with crocuses.

♨ Temple of Arethusa J12

Designed by Sir William Chambers, and built in 1758, this temple is close to the Campanile. It contains memorial tablets to members of Kew staff who lost their lives in the two World Wars.

♨ Ruined Arch L7

This arch was designed as a ruin by Sir William Chambers and built in 1759–60, when mock antique buildings were fashionable.

♨ Japanese Gateway or Chokushi-Mon J3

The Japanese Gateway is situated on a small mound not far from the Pagoda and close to Cedar Vista. It is a replica, four-fifths natural size, of a

famous Japanese Gate 'The Gateway of the Imperial Messenger' or Chokushi-Mon, which was one of the exhibits at the Japanese-British Exhibition at Shepherd's Bush in 1910. It was presented to the Gardens by the Kyoto Exhibitors' Association at the end of the exhibition.

⚲ Flagstaff K8

The Flagstaff, which is the third to be erected at Kew since 1861, is 68.5 m (225 ft) high and was made from a single trunk of Douglas Fir, *Pseudotsuga menziesii*, which was about 371 years old when felled. The tree came from Copper Canyon, near Chemainus, Vancouver Island, and was presented by the Government and Forestry Industry of British Columbia. It was put up by the 23rd Field Squadron, Royal Engineers on the 5th November 1959.

⚲ Queen's Beasts H/J12

The ten Queen's Beasts in front of the Palm House are replicas in Portland Stone of the ones which stood outside Westminster Abbey at the time of the Coronation in 1953. They are by the same sculptor, the late Mr James Woodford OBE, RA, and were presented to Kew anonymously in 1956.

⚲ Campanile K12

The Campanile to the north of Victoria Gate was built in the 1840's and is just over 32 m (107 ft) high. It is connected to the Palm House by a tunnel and was designed both as a water tower and (unsuccessfully) as a chimney for the boilers which were originally under the Palm House.

⚲ Ice House H15

In the seventeenth, eighteenth and early nineteenth centuries, ice was harvested in winter, and stored in ice houses packed with straw to aid insulation. The ice was used mainly for cooling drinks and keeping food fresh in summer. It is not known exactly when this ice house was built, but it probably pre-dates the founding of the Botanic Garden in 1759. The house was restored in 1982.

⚲ Museum I (GENERAL MUSEUM) J/K13

This Museum displays a range of economic (useful) plants including those used for dyes, medicines, oils and waxes, rubber and poisoned arrows, etc. Near the entrance is a model indigo factory of Nuddea, India, and a moss box showing a range of British mosses, liverworts and other cryptogams can also be seen here. The ground floor only is open to the public.

◀ **The Chokushi-Mon (Japanese Gateway) with Japanese Azaleas**

The building, opened in 1857, was designed for use as a museum by Decimus Burton, as a result of the popularity of the world's first museum of economic botany which was opened at Kew in 1847, and housed in a converted gardener's dwelling and fruit store. This original building is now closed to the public, but may be seen opposite the Aquatic Garden (J16).

♨ Wood Museum H17

The Wood Museum is housed in part of Cambridge Cottage, which was presented to the Gardens by King Edward VII after the death of the last Duke of Cambridge in 1904. It is approached through Cambridge Cottage Garden. The interior of the building has been remodelled to make it suitable for a museum, but the outside is unchanged except for the removal of a narrow, first-floor balcony. Woods from British-grown trees (though not necessarily native species) are displayed along a corridor, with softwoods, from gymnosperm trees (mostly conifers), to one side, and hardwoods, from broad-leaved trees, to the other side. There are also some very fine examples of inlay work and a variety of wooden objects including a cannibal's club and eating bowl.

The first floor is devoted to past and present uses of timber including paper, chipboard and hardboard manufacture.

♨ Marianne North Gallery L7

The Marianne North Gallery is situated between the Flagstaff and Refreshment Pavilion. It houses a unique collection of 832 oil paintings by Miss North, who gave them to Kew, together with the building specifically designed by her architect friend James Fergusson. The paintings are of plants, insects and general scenes from the many countries she visited on her world-wide travels between 1871–85. They were arranged in the gallery by Miss North and the whole effect is breath-taking.

♨ Orangery F16

The Orangery is now an exhibition area housing the bookshop, a temporary exhibit of current interest and a display giving a brief history of the Gardens and the functions of the various Divisions of Kew including sections not open to the public.

Designed by Sir William Chambers, and built in 1761, it was used as an orangery until 1841, then as a glasshouse, although lack of light meant the plants barely survived. It later functioned as a wood museum. Attempts to restore the building to an orangery in 1959 were not wholly satisfactory and so it was turned to its present use.

◀ The Marianne North Gallery

♨ Kew Palace D16

Kew Palace, or the Dutch House, was built in 1631 for Samuel Fortrey, a merchant of Dutch origin, on the foundations of a sixteenth-century building, called the Dairy House. The brick is laid in Flemish bond (with the sides and ends alternating), which was then a new fashion in England. The initials of Samuel Fortrey and his wife Catherine can be seen over the main entrance.

George III acquired this building in the eighteenth century and used it as an annexe to Kew House which lay to the south of the Palace (the site is marked by a sundial), particularly for some of his 15 children. In 1802, Kew House was pulled down and the Palace was then used regularly by the royal family. In 1818 the marriages of the Duke of Clarence (later King William IV) and the Duke of Kent (father of Queen Victoria) took place in the drawing room. Four months later, Queen Charlotte died in the house which was then shut up until 1899, when it was opened as a museum. Today it is administered by the Department of the Environment and is open to visitors in the summer months.

♨ Queen Charlotte's Cottage and grounds E2

The Queen's Cottage was built about 1772 as a picnic place for Queen Charlotte, wife of George III, and their family. It was designed for appearances rather than practicality, and was never intended for habitation. The building, administered by the Department of the Environment, is open to visitors on summer weekends and Public Holidays (from April to September).

The Queen's Cottage Grounds were the last main addition to the Gardens. Situated in the south-west corner of the Gardens, they comprise about 16 hectares (40 acres) of woodland thickly carpeted with bluebells in May. The area was presented by Queen Victoria to commemorate her Diamond Jubilee, and first opened to the public on 1st May 1899. Access is restricted to fenced paths, because a condition of the gift was that it should remain in its semi-natural state. In the late nineteenth century, 70 birds species were listed for this area, including woodcock, nightingale and kingfisher. It is still possible to see a wide variety of birds here, often at close quarters.

Bluebells also flourish by the Woodland Walk, a semi-wild area to the west of the Temperate House. The Bluebell, *Hyacinthoides nonscripta*, is occasionally confused in name with the Bluebell of Scotland or Harebell, *Campanula rotundifolia*, an unrelated plant found on dry open pastures and heaths.

Herbaceous and formal gardens

♨ Herbaceous Ground J15, 16

This area was a walled kitchen garden for the royal estate at Kew until
1846, when it was incorporated into the Botanic Gardens. Since then,
annual and perennial herbaceous flowering plants have been grown here.

The main collection is of temperate dicotyledons, with some
monocotyledons planted near the walls. Plants are arranged in their
families following the system of Bentham and Hooker. Related plants are
thus near to one another, allowing easy comparison for botanical study.
The collection contains approximately 2000 species, of which about 50 per
cent are of wild origin. A pergola covered with climbing roses transects the
ground; assorted ivies grow on the east wall, and various shrubs, climbers
and trees against the west wall.

♨ Grass Garden H/J16 and Bamboo Garden C/D9

The formally arranged Grass Garden, situated at the northern end of the
Rock Garden, contains approximately six hundred different members of
the Grass Family (*Gramineae*), and displays their diversity and usefulness.
The peripheral beds are symmetrically arranged for effects of flower, fruit
and leaf showing the architectural beauty and variety of these plants, with
focal points provided by clumps of pampas grasses (*Cortaderia* species and
cultivars).

The central beds, which display temperate and tropical cereals, are set
each side of a vista terminated by the bronze statue of 'A Sower' by Sir
Hamo Thorneycroft R.A. presented to Kew in 1929. Flanking the statue
are sample 'plots' of lawn grasses.

Bamboos are woody grasses mostly from warm regions. Although a few
are in the Grass Garden, the main collection is in the Bamboo Garden,
east of the Rhododendron Dell. It was constructed in 1892 and was
probably a factor responsible for popularising bamboos in other gardens at
that time. The sunken area provides the shelter and moist soil required by
bamboos.

♨ Rose Garden G/H11, 12

The Rose Garden lies on the south and west sides of the Palm House and
is formal in design. Large flowered (Hybrid Tea) and Cluster Flowered
(Floribunda) roses are grown, both old and new varieties, and there is a
collection of 'old roses' which traces the ancestry of the modern rose
groups.

William Nesfield, the nineteenth-century landscape architect, used the

Palm House as a focal point for the vistas he designed to give interest to an otherwise dull area. Standing outside the western central door a good view of all three vistas can be obtained–straight ahead is Syon Vista which runs down to the river, on the left is Pagoda Vista, and on the right the third one stretches down to the big cedar near the Beech Clump.

🜲 Palm House bedding (H/J12) and Broad Walk E16–H13
The beds in front of the Palm House are a modification of the original Victorian parterres designed by W. A. Nesfield. Colourful and sometimes unusual bedding schemes both here and along the Broad Walk are at their best in spring (April–May) and summer (June–September).

🜲 Rock Garden J15, 16
The Rock Garden was originally laid out in 1882 as an alpine valley, the main path representing the course of a river, with waterfalls and streams at the sides. The rock used at first was hard limestone, but this was found to reflect too much heat and did not hold enough water, so over a period of years it was replaced with Sussex sandstone.

The Rock Garden is at its best in spring, though there is something of interest to be seen at most seasons of the year. Popular and rare montane plants from many different countries are grown, including bulbs, small perennials, dwarf conifers and shrubs.

🜲 Duchess's Border H/J17
Just north of the Grass Garden, against the south-facing wall of Cambridge Cottage Garden, there is a narrow but interesting border of tender plants from places such as New Zealand, Tasmania, Mexico and the Mediterranean. Notable is an Olive, *Olea europaea*, and a very large *Sophora microphylla* which occurs in Chile and New Zealand and is covered each April in a mass of spectacular yellow blossom.

Growing actually from the wall are four *Thuja orientalis*, the Chinese *Arbor-Vitae*, which are thought to have been planted there. These hardy conifers grow very slowly and date at least from the nineteenth century. A branch from one of the Thujas, removed in 1982, had 90 annual rings!

🜲 Cambridge Cottage Garden H/J17
This charming walled garden is also known as the Duke's Garden, after the Duke of Cambridge who lived in Cambridge Cottage. A collection of the more ornamental herbaceous plants are grown here in the side borders and island beds in the main lawn. In the far corner a large area of paving has been used to house a collection of bulbs and this is backed by a border of plants demonstrating the different types of variegation. The bulbs also

have an historical link in that the present Queen's Garden (D16) is on a site known in the late nineteenth century as the Duke of Cambridge's bulb garden. Some species irises are also displayed here and around the sheltered footings of the 'cottage' are several different forms of the beautiful winter flowering *Iris unguicularis* which comes from the Mediterranean region.

Iris enthusiasts will also wish to note the moisture-loving irises that are grown in the Woodland and Rock Gardens, and the special raised beds made for Juno, Oncocyclus and Regelia irises near the Jodrell Laboratory – which also hosts some rare bearded irises against its south face.

Trees grown here include the Iron Tree, *Parrotia persica*, native to Persia and the Caucasus. It is a member of the Witch Hazel family, *Hamamelidaceae*, and has dark crimson flowers which appear from mid-January to early March, and later in the year the tree produces good autumn colour. *Eucommia ulmoides*, an unusual tree from Central China, is interesting on account of the rubber produced in its leaves (the only tree hardy in this country to do so); in China it is also valued for its bark, which is believed to have medicinal properties.

A number of magnolias may be seen including several specimens of the evergreen *Magnolia grandiflora*, native to the southern United States, which produces deliciously scented large creamy white flowers in late summer and early autumn.

⚘ The Queen's Garden D16

This garden is behind Kew Palace and was opened by Her Majesty the Queen in May 1969. It is laid out in the style of a seventeenth-century garden and contains appropriate features such as a parterre, sunken garden, gazebo, pleached alley and a 'mount' with a rotunda on top. Only plants which would have been grown at that time are included. On many of the labels there are amusing quotations from early herbals, showing the use to which the plants were put. Various sweet-smelling herbs are grown, as they were extremely important 300 years ago, being used to disguise the flavour of bad meat and to mask the unpleasant smells which abounded both indoors and out. They were also frequently used as medicines.

⚘ Aeolus Mound and Woodland Garden J14

In spring the slopes of the mound are garlanded with snowdrops, daffodils and wild flowers until late June when the bulb leaves have withered and the long grass is cut. On its northern side, under the shade of young oak trees, nursed by birch, there is a collection of woodland herbs set amongst many fine rhododendrons. Most noticeable in season are the hostas, lilies and primulas, but this area holds many other floral delights.

🔥 Economic Plants Garden E17

Begun in 1984, as part of the Sir Joseph Banks Building, a new 5 acre garden is being developed behind House No. 1. The plantings will consist of trees, shrubs, and herbaceous plants of economic use, past and present. The surrounding landscape will relate to the Building with its theme of the usefulness of plants for mankind.

🔥 Heather Garden L4

The Heather Garden is a landscaped area approached from either the Pagoda, Lion Gate, or the eastern boundary path. The main flowering periods for heaths and heathers are September–October and February–March, but colour is also given throughout the year by variegated foliage. Height is provided by small conifers including junipers, and clumps of Tree Heath, *Erica arborea*. It is from the woody rootstocks of this plant that briar pipes are made, the name being derived from bruyère – French for heath or heather. *Gaultheria shallon*, another member of the Heather family, is a native of N. America. It produces edible berries and is planted in Britain as cover for game. A few plants from other families with heath-like flowers and foliage have also been planted here.

Trees and Shrubs

🔥 Birch collection E13

The birch collection, just south of the Tea Bar and on towards the Rhododendron Dell, is easily recognised by the white trunks of most species. However, the River Birch, *Betula nigra*, has an attractive chocolate-brown flaky bark and the Sweet Birch, *B. lenta*, is also dark-coloured. Birches are fast-growing but short-lived trees found in temperate areas of the northern hemisphere. Some flourish in arctic-alpine conditions, where they are valuable for firewood. The bark is often more durable than the wood and its waterproof nature allows it to be used for roofing, canoes and even cups. The Paper Birch, *B. papyrifera*, has been used in this way by N. American Indians.

🔥 Poplar and willow collections C13, D12, 13, E7, 8, F8

The poplar collection is found near the birch collection between Brentford Ferry Gate and Princess Walk. Poplars are deciduous trees native to the north temperate zone and are very fast-growing, especially on heavy or wet soils. They have vigorous, wide-spreading roots which are often responsible for damage to drains and foundations of buildings. The timber

of certain species is used to make matches, chip-baskets, plywood, etc. The Aspen, *Populus tremula*, a British species growing beside the Lake, has leaves which quiver even in the slightest breeze and turn a wonderful yellow in autumn. The buds of Black Cottonwood, *P. trichocarpa*, one of the balsam poplars from N. America, give a delicious scent to the air.

Willows thrive in damp positions. They range from creeping shrubs to large trees and are found principally in north temperate and arctic regions. Weeping forms and those with a bushy habit are grown by the Lake (E7,8–F8). Some produce useful timber, such as *Salix alba* var. *caerulea* which provides wood for cricket bats. Others, particularly the larger shrubby kinds known as osiers, are cut back regularly to provide pliant shoots for basket-making.

♨ Oak collection C7–10

The oak collection extends from Mount Pleasant north-east along River Side Avenue. Of about 300 *Quercus* species from the north temperate zone and high altitude tropics, about 70 species will grow in our climate. All bear acorns in natural conditions, but may be trees or shrubs, evergreen or deciduous.

South of the Rhododendron Dell and between the two Avenues is a Kermes Oak, *Q. coccifera*, a 3 m (10ft) shrub with small, holly-like leaves. Native of the Mediterranean region, it was once important as the host plant of the Kermes insects which provided a scarlet dye used in making cardinal's 'purple' in the Middle Ages.

Other oaks are valued for their timber, especially the British oaks, *Q. robur* and *Q. petraea*. The bark of the Cork Oak, *Q. suber*, and Chinese Cork Oak, *Q. variabilis*, provide cork, and the brilliant autumn tints of the Scarlet and Red Oaks of N. America, *Q. coccinea* and *Q. rubra*, are a splendid sight.

Other notable and older specimens are to be found throughout the Gardens:

The Chestnut-leafed Oak, *Q. castaneifolia* (G13), is probably the most magnificent tree in Kew and is the oldest specimen of its kind in this country, planted in 1846.

Two hybrid oaks have also formed large trees; in each case one parent is deciduous and the other evergreen. The Lucombe Oak, *Q.* × *hispanica* 'Lucombeana' (G10), is a hybrid between the Turkey and Cork Oaks. Turner's Oak, *Q.* × *turneri* (G15), is a hybrid between the English and Holm Oaks.

Two specimens of the Sessile or Durmast Oak, *Q. petraea*, were grown from acorns collected on the battlefield of Verdun in 1916. One is near the Chinese Guardian Lions (J12) and the other on the mound (J13).

The Holm Oak, *Q. ilex*, is probably the best-known evergreen oak, and was introduced from the Mediterranean in about 1500. Specimens line Syon Vista (E8, 9) and are common elsewhere in the Gardens.

♨ Alder collection F9

The alders are deciduous trees and shrubs found chiefly in north temperate regions and southwards to Peru and W. China. Most species do best in damp places, so the main collection has been arranged around the eastern end of the Lake where there is plenty of moisture. The flowers are borne in catkins produced in early spring, the male being long and showy and the female quite small and erect, ripening to persistent cone-like structures. *Alnus glutinosa* is the only species native to this country. The timber was formerly used to make the soles of clogs; today it is used in turnery – for broom backs, tool handles, etc. *A. glutinosa* 'Imperialis' has the leaves deeply cut, sometimes almost to the midrib. *A. subcordata* from the Caucasus and Iran has large leaves which often do not fall until the end of November. Another handsome species is *A. cordata* from Corsica and S. Italy, which was introduced in 1820 and forms a pyramidal-shaped tree.

♨ Ash collection F10 TO F13

The ash collection is on both sides of Princess Walk interplanted with other members of the Olive family, *Oleaceae*. The shrubs such as privets, *Ligustrum*, and jasmines, *Jasminum*, are in island beds. There are about 60 species of ash, *Fraxinus*, mostly deciduous trees in the north temperate region. Many are represented here and most have the characteristic pinnate leaves. There are also many varieties and forms of the Common ash, *Fraxinus excelsior*. This native tree, often reaching 30 m (100 ft), has tough, elastic timber previously used for spear shafts and coachwork and now employed in sports goods such as hockey sticks and tennis rackets. The several specimens of Manna Ash, *F. ornus*, produce striking white flowers in May. In the past, a sugary exudate (manna) from incisions in the bark of this tree was used as a laxative. This is not, however, the 'manna' of the Bible.

♨ Conifer collections B5 to G3 and D5 to F8

The main conifer collections are along Boathouse Walk and on either side of the path leading from Isleworth Gate to the crossroads near Oxenhouse Gate. For many years these trees suffered badly from the effects of air pollution, but since London became a smokeless zone their growth and general condition has improved.

Because of adverse conditions, it was decided in 1923 to set up jointly

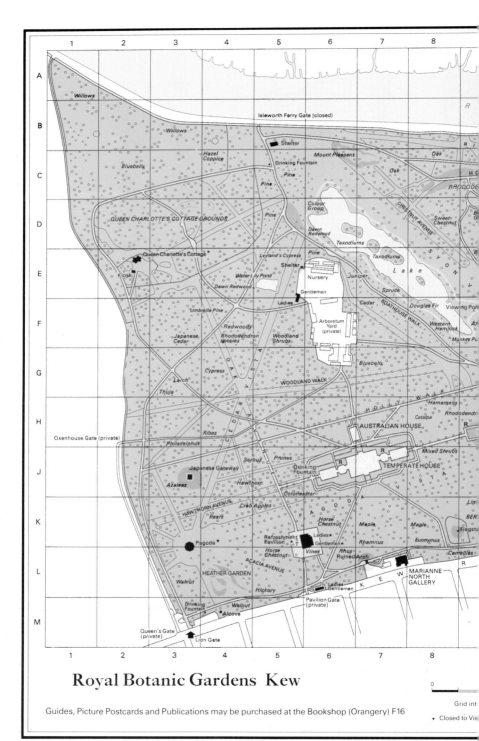

Royal Botanic Gardens Kew

Guides, Picture Postcards and Publications may be purchased at the Bookshop (Orangery) F16

Grid int

▾ Closed to Vis

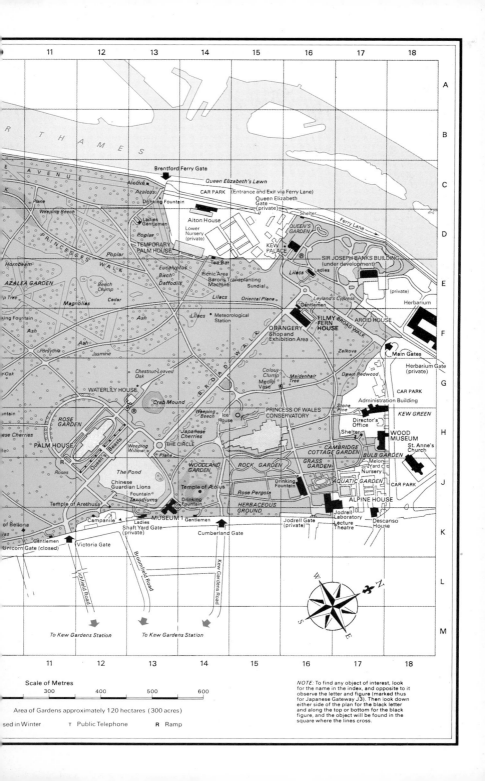

with the Forestry Commission, a National Pinetum at Bedgebury in Kent, in order that a comprehensive collection of conifers could be established in an unpolluted area. Bedgebury is now an independent Pinetum but Kew's collections have been extended by the addition of Wakehurst (p. 6), whose moist climate especially favours *Abies, Picea, Tsuga*, etc.

Starting from the east end of Boathouse Walk, it will be found that nearly all the species on the right are from the New World and those on the left are from the Old World. The first trees to be encountered are the Monkey Puzzles, *Araucaria araucana*, which were introduced to this country from Chile in 1795. Species of *Tsuga* may be seen next, including the Western Hemlock, *T. heterophylla*, from N. America, which is a valuable source of pulpwood. There are several specimens of the Douglas Fir, *Pseudotsuga menziesii*, an important softwood, first discovered by Archibald Menzies on Vancouver Island in 1791 and subsequently introduced to Britain by David Douglas in 1827. These are followed by the silver firs, *Abies*, and the spruces, *Picea*. The Norway Spruce, *P. abies*, is familiar for its use as a Christmas tree and *P. sitchensis* is now of great importance in British forestry. The junipers include *Juniperus communis*, one of our three native conifers, whose fruits are employed to flavour gin, and *J. virginiana*, well-known for its fragrant wood used to make pencils. Several species of *Juniper* may also be found in the Heather Garden (L4).

The pine collection stretches from the west end of the Lake to Isleworth Gate. The pines are of great economic importance, providing valuable timber, turpentine and rosin. Their needles are nearly always arranged in bundles of 2, 3 or 5, which is a useful aid to identification. The Scots Pine, *Pinus sylvestris*, another of our native conifers, is a 2-needle pine. The Monterey Pine, *P. radiata*, has a distinctive thick fissured bark, while the Lacebark Pine, *P. bungeana*, has a most beatuiful bark, which scales off leaving a jigsaw-like pattern of many different colours. Pines with edible seeds include the Stone Pine, *P. pinea*, and the Arolla Pine, *P. cembra. P. rigida* produces clusters of cones that persist for many years, while the cones of the Bishop Pine, *P. muricata*, may remain on the tree for 40 years or more.

Taking the path from the Lake to the crossroads near Oxenhouse Gate, the rest of the collection can be studied. There are two fine specimens of the Leyland Cypress, × *Cupressocyparis leylandii*, a hybrid between *Cupressus macrocarpa* and *Chamaecyparis nootkatensis*. In the Waterlily Pond (E5) there is a good example of the deciduous Swamp Cypress, *Taxodium distichum*, from the Mississippi area, showing its breathing roots or 'knees' sticking up around the base of the trunk (better examples of the 'knees' can be seen round the Palm House Pond). Another interesting deciduous conifer is the Dawn Redwood, *Metasequoia glyptostroboides*, which was first

◀ **Stone Pine, *Pinus pinea***

discovered in China in 1941, having been thought previously to have been extinct for over 100 million years. The Maidenhair Tree, *Ginkgo biloba*, is another primitive tree native to China, which has been called a 'living fossil', as it is the only representative of a group of trees which flourished about 180 million years ago. Nearby are specimens of *Cephalotaxus fortunei*, introduced from China by Robert Fortune in 1848, and known by the delightful name of Chinese Cow's Tail Pine. Farther on are the related *Podocarpus* species, including the Plum-fruited Yew, *P. andinus*, and opposite them the true yews, *Taxus*, including many forms of our native Common Yew, *T. baccata* – one particularly attractive variety has yellow 'berries'.

A number of unusual conifers then follows on the south side of the path (in this area the Old World species are mainly on the right and New World on the left). These include *Torreya* species, known as Nutmeg Trees from the resemblance of their seeds to this spice, the Chinese Fir, *Cunninghamia lanceolata*, with red stringy bark whose timber is valued as coffin wood, the Japanese Cedar, *Cryptomeria japonica*, an important timber tree in Japan, and the Umbrella Pine, *Sciadopitys verticillata*, also native to Japan, whose leaves consist of 2 fused needles.

Opposite, there is a grove of the famous redwoods of California. The Coastal Redwood, *Sequoia sempervirens*, attains a great height in its native country; one (from which a seedling is growing in this collection) has the distinction of being the tallest tree in the world, 112.4 m (369 ft) high. In this country the tallest are just over 40 m (131 ft). The Giant Redwood, Big Tree or Wellingtonia, *Sequoiadendron giganteum*, will also grow to a great height, but it is for its immense girth, up to 25 m (82 ft) that it is so famous; the oldest specimens in California are believed to be about 4600 years old.

Beyond them various species of *Cupressus* and *Chamaecyparis*, including *Chamaecyparis lawsoniana* and its numerous cultivars, can be seen. Because of their deciduous habit the larches, *Larix*, survive here better than many of the evergreen species; they can be seen on either side of the path leading to the Temperate House. In autumn the leaves of the related Golden Larch, *Pseudolarix amabilis*, turn gold, orange and finally brown. The last genus encountered is *Thuya*, which includes the Western Red Cedar, *T. plicata*, one of the most useful timber trees of western N. America.

♨ *Rosaceae* collection J/K4, 5

The woody *Rosaceae* collection is found between Cedar Vista and the Temperate House. Most common fruits of the north temperate zone such as apples, pears, plums, quinces, cherries, blackberries and raspberries belong to this family, as do many ornamental trees and shrubs including

◀ Spring scene – daffodils and flowering cherries

flowering cherries, cotoneasters, pyracanthas, mountain ashes (*Sorbus*), spiraeas, hawthorns (*Crataegus*), etc. Many ornamental crab-apples can be seen on the mound near the Waterlily House (G13). Herbaceous members of this family are to be found in the Herbaceous Ground. (J15).

🐿 Walnut collection L3–5, M4

Near the Lion Gate a number of different species of walnut can be seen. The walnuts are native to N. and S. America, Europe and Asia, and of the 15 species known about 8 or 9 are in cultivation. The Common Walnut, *Juglans regia*, native to S.E. Europe eastwards to China, was introduced to this country in early times and is well known for its nuts and valuable timber. The Black Walnut, *J. nigra*, from eastern N. America has such hard-shelled nuts that special nutcrackers have to be made to open them. The fallen fruits of several species smell of scented soap.

The hickories, *Carya* species, belong to the Walnut family, and with one exception are all native to eastern N. America. The Shagbark Hickory, *Carya ovata*, produces edible nuts which are highly valued in the USA. The Pignut Hickory, *C. glabra*, which is rare in cultivation in this country, has a most attractive fissured bark. There is also a fine specimen of the Mockernut Hickory, *C. tomentosa*.

The wingnuts, *Pterocarya* species, also of this family, get their common name from their fruits which are surrounded by green wings and borne in long catkins. *P. fraxinifolia* from the Caucasus was introduced in 1782 and *P. stenoptera* from China in 1860; their hybrid *P. × rehderana* can be seen here and elsewhere in the Gardens.

🐿 *Leguminosae* collection L4, 5

The collection of trees belonging to the *Leguminosae*, (Pea family), is situated between the Heather Garden and the Refreshment Pavilion. Running from the Pavilion towards the Pagoda is Acacia Avenue, which, like streets of that name, is planted with varieties of the False Acacia or Locust Tree, *Robinia pseudacacia*, as well as the Honey Locust, *Gleditsia triacanthos*, many specimens of which are remarkable for their fierce spines.

Shrubby *Leguminosae* such as gorse, *Ulex europaeus*, Common Broom, *Cytisus scoparius*, and the closely related *Genista*, are planted around King Willam's Temple (H9). *Genista tinctoria*, Dyer's Greenweed, was once used to dye cloth, while gorse, also known as furze or whin, was used for firing bakers' ovens in country areas.

Other specimens in the Gardens include two old trees planted in the original 9 acres of the Botanic Garden soon after it was established in about 1759. One is a False Acacia, *Robinia pseudacacia*, a native of N.

America, growing to the south-east of the Orangery near the Maidenhair Tree (G16), and the other a Pagoda Tree, *Sophora japonica*, of which only two recumbent side branches remain; it is next to the path (G15).

An interesting curiosity is + *Laburnocytisus adamii*, a graft hybrid between laburnum and Purple Broom. A bed devoted to this tree and the two 'parents' can be seen near the crossroads (G17). (See also Wisteria, p. 39).

🎋 Berberis Dell K9

The Berberis Dell is behind the Temple of Bellona. In the centre of the hollow and to the west near the Flagstaff are the barberries, *Berberis* species, and the closely related mahonias which differ from *Berberis* in having pinnate leaves and thornless shoots. On the north slope of the Dell is a bed of hardy hibiscus together with plants of the Ivy family, *Araliaceae*. Hardy exotic trees, including *Eucalyptus gunnii* (see p. 38), border the footpath to the east.

🎋 *Catalpa* collection H8

Catalpa species from both the Old and New World are represented in the Kew collection, and the best known is probably *Catalpa bignonioides*, the Indian Bean. This native of the eastern USA is one of the last trees to leaf, in late June, rapidly followed by the large white flowers in mid-July. These in turn produce long slender pods which, despite their appearance and the common name of the tree, are not edible. Other specimens can be found throughout the Gardens.

🎋 Mulberry collection H9

The mulberry trees, *Morus*, grow to the north of King William's Temple. The collection includes a small Common or Black Mulberry *Morus nigra*, but a larger specimen is to be found about 18 m (20 yd) west of the midpoint of the Broad Walk. This slow-growing tree gives old-world charm to any garden, and produces blackish-red fruits which are really clusters of drupes, each from a minute flower. The leaves of the White Mulberry, *M. alba*, are the food of silkworms, and in the East large numbers of trees, often coppiced, are grown for silk production. The fruits are used medicinally, the bark for paper-making and the wood is excellent for sports goods. Across the path is *M. kagayamae*, which probably grew from seed collected in Japan in 1917 by E. H. Wilson.

🎋 Rhododendrons and Azaleas

The rhododendrons and azaleas belong to the Heather family, *Ericaceae*, and are now both included in the genus *Rhododendron*. The name azalea is

popularly applied to those with deciduous leaves while rhododendron is used for the evergreen species.

One of the most attractive areas planted with rhododendrons is the Rhododendron Dell (c8–11), which was designed by 'Capability' Brown in the latter part of the eighteenth century and then known as the Hollow Walk. With the introduction of many rhododendrons in the mid-nineteenth century, particularly ones from the Himalayas collected by Joseph Hooker, this area was planted up as a Himalayan rhododendron valley. Although it retains that character and some of the original plantings still remain, today many of the plants are hybrids of garden origin and not specifically Himalayan.

The main *Rhododendron* species collection is found on either side of Cedar Vista (F–G4–5), where they are grown among the trees and underplanted with primulas, Summer Snowflakes, hostas, etc. Another particularly noteworthy area for rhododendrons is the Woodland Garden between the southern end of the Rock Garden and the Mound (J14). A major collection of *Rhododendron* species is now also being established at Wakehurst Place.

The Azalea Garden (E10) was laid out over 100 years ago and is at its best in late May and June. It is also attractive in the autumn with the leaves turning shades of red and orange. Japanese azaleas can be seen on the western and southern slopes around the Japanese Gateway (J3).

☙ *Magnolia* and *Liriodendron*

Many species of *Magnolia* can be seen flowering in spring, particularly near the Temple of Bellona (K11), and near the Beech Clump (E12). The genus is named after Pierre Magnol (1638–1725), a professor of botany and medicine at Montpellier. One of the earliest to flower is *Magnolia campbellii*, native to the Himalayas, which produces enormous pink flowers, the size of soup plates. Another early-flowering species, *M. stellata*, from Japan, bears beautiful white star-like flowers; it is a good magnolia for the small garden as it flowers when quite young. *M. × soulangiana*, perhaps the most popular of all magnolias, is a hybrid between two Chinese species, *M. heptapeta* and *M. quinquepeta*; it was first raised in the garden of Soulange-Bodin near Paris in 1826.

The Tulip Tree, *Liriodendron tulipifera*, also belongs to the Magnolia family and was one of the earliest introductions from N. America. It can be recognised by its unique leaf, which appears to have the end cut off. The flowers are orange and greenish in colour and when young have a delicious smell of vanilla; however, they are seldom produced until the tree is 30 years old or more. Tulip Trees may be seen on either side of the Broad Walk and there is also a fine specimen in the Azalea Garden. *L. chinense*

◀ **The Rhododendron Dell**

(F12), the only other species in the genus, was introduced from China in 1901.

♨ Eucalyptus

Virtually all eucalypts (gum trees) are native to Australia and Tasmania where they dominate the forest areas. Of about 600 species, only the Cider Gum, *Eucalyptus gunnii*, is truly hardy in eastern Britain, and a fine specimen over 18 m (60 ft) tall can be found just west of the path, halfway between the Flagstaff and Victoria Gate (K10).

Other less hardy eucalypts are planted to the north and north-east of the Aquatic Garden (J16), south west of the Tea Bar (D, E13), and just north of the Ruined Arch (L7). On the younger trees or where there has been drastic pruning, you can see the juvenile leaves, rounded and often clasping the stem. These are popular with flower arrangers. The longer, thinner, adult leaves seen on the older tree often hang downwards, giving rise to the term 'shadeless forests'. Eucalyptus oil, frequently used in throat lozenges, is distilled from young shoots of *E. globulus* and other species, including *E. gunnii*. The flowers are the main source of pollen and nectar for Australian bees, and the leaves of some species provide food for Koala Bears.

Some eucalypts are grown in the Australian House (p. 48).

♨ Arbutus – Strawberry Trees

The Strawberry Trees belong to the Heather family, *Ericaceae*, and are natives of N. and C. America, the Mediterranean and S. W. Ireland. *Arbutus unedo* produces globular red strawberry-like fruits, which are not very tasty; the specific name *unedo* signifies that to eat one fruit is enough. In some countries jellies and other preserves and an alcoholic drink are prepared from the fruits. Near King William's Temple (H9), *A. unedo* and *A. andrachne* from S. E. Europe may be seen with their hybrid *A.* × *andrachnoides*. Notice their beautiful red bark.

♨ Davidia involucrata var. vilmoriniana – Handkerchief Tree E11, H9

This tree is named after Père David, a French missionary, who first discovered it in China in 1869. The variety *vilmoriniana* is commoner in cultivation than the type, which was introduced from W. China in 1904. The common names Handkerchief Tree, Dove Tree and Ghost Tree refer to the large showy white bracts which hang down from the globular, dark purple-red flowerheads.

🌿 Lilacs

The lilac collection consists mainly of cultivars of the Common Lilac, *Syringa vulgaris*, and is in two areas: in front of Kew Palace (E16) and south-east of the Tea Bar (E, F14). The Common Lilac, now widely grown for its beauty and fragrance, was first brought to this country from E. Europe at the end of the sixteenth century and by 1601 was well established. In some districts of France a decoction of lilac bark was used to ease fevers and agues.

Mock-orange or Syringa (*Philadelphus*) is no way related to lilac. The confusion of *Syringa* being the botanical name for lilac but the common name for *Philadelphus* was started by John Gerard in his herbal of 1597, where he called both plants *Syringa*.

🌿 Wisteria

The wisterias are a small group of ornamental, deciduous, climbing shrubs of the *Leguminosae* family, with beautiful blue, lilac or white flowers produced from May–August. They are native to China, Japan and USA, but the Kew collection contains only oriental species.

Wisteria sinensis from N. China was the first species to be introduced to England in 1816, from the garden of a Cantonese merchant. Soon after, plants sold for 6 guineas (£6.30) each, but by 1835 cost only 1s.6d (7p). A very old specimen grows over an iron cage close to the Maidenhair Tree (G16). Its age is unknown, but it was growing against Princess Augusta's 'Great Stove' when that was demolished in 1861, and it was then that the cage was built. A white form grows against the wall in the Herbaceous Ground (J15).

Wisteria floribunda forma *macrobotrys* is a cultivated form of a Japanese species. It has longer racemes of flowers than *W. sinensis* and is less suited to growing against walls. One famous specimen in Japan forms a huge arbour and produces thousands of pendulous racemes each often exceeding 1 m in length. A clump grows just west of the Stone Pine (G17) and also against the wall facing the Rock Garden (J16). *W. floribunda* forma *alba*, to the east of the Stone Pine (H17) and south-west of the Waterlily House (G12), has smaller racemes of white flowers.

🌿 Forsythia

The forsythias belong to the Olive family, *Oleaceae*, and with the exception of one rare relict species, *F. europaea*, from S.E. Europe, all are native to E. Asia. The genus is named after William Forsyth (1737–1804), who was superintendent of the Royal Gardens at Kensington. A collection of these shrubs can be seen near the ash collection and elsewhere in the Gardens.

Winter-flowering shrubs

🌢 *Chimonanthus praecox* – Winter Sweet
This native of China is a shrub best grown against a wall. Its merit is in the small yellow-white waxy flowers produced usually in December but which may occur any time between November and March. They have an extremely strong scent and it is said that one flower can perfume a room. Specimens can be seen on Cambridge Cottage Garden wall (H17) and just south of the Ruined Arch (L7).

🌢 *Cornus mas* – Cornelian Cherry
The Cornelian Cherry is native to Europe and has been cultivated for centuries in Britain. Before the introduction of the oriental witch hazels, it was the only effective yellow winter-flowering shrub or tree. It produces its flowers before the leaves in early February. The fruits are red and were formerly made into syrups and preserves; they are occasionally produced in this country. Good examples can be seen on the west side of the Rock Garden (H15), behind the Orangery, and near King William's Temple.

🌢 *Hamamelis* – Witch Hazels
The main collection of witch hazels is situated near King William's Temple (H9), but many specimens can be seen in other parts of the Gardens. *Hamamelis mollis* from China and *H. japonica* from Japan are some of our most beautiful winter-flowering shrubs; cultivars, and the hybrid *H. × intermedia* and its cultivars, are also grown. The least showy species is the autumn-flowering *H. virginiana* which was introduced from N. America in 1736. Various remedies are prepared from decoctions and extracts of the leaves and young twigs and it is still cultivated in this country for this purpose.

🌢 *Prunus subhirtella* 'Autumnalis'
This winter cherry flowers from October to Christmas and often again in April. The flowers are semi-double and white or pale pink in colour. Several examples can be seen near the Pagoda.

🌢 *Viburnum farreri (V. fragrans)*
This attractive shrub, introduced from N. China in 1909, has deliciously scented flowers which are produced from November onwards. The hybrid between this species and *V. grandiflorum* from the Himalayas is *V. × bodnantense*; the variety 'Dawn' is a popular form. Specimens can be seen near the Orangery and near Victoria Gate.

Greenhouses

🌿 Aroid House (NO. 1 HOUSE) E/F17

Originally one of a pair of greenhouses designed by John Nash for the gardens of Buckingham Palace, the Aroid House was moved piecemeal to Kew in 1836 on the orders of King William IV.

The house is kept at a minimum temperature of 18.3°C (65°F) with high humidity and in summer the roof is shaded. It is therefore an ideal place for growing plants from the lower levels of tropical rainforests, such as epiphytes and lianes. Epiphytes (*epi*=upon, *phyton*=a plant) grow on other plants (but do not take nourishment from them) and produce long aerial roots which are able to take moisture directly from the air and sometimes reach the ground and enter the soil. Lianes (or climbers), which climb high up into the trees, also produce long aerial roots. Many of the plants grown are aroids (i.e. belonging to the monocotyledonous Arum Lily family, *Araceae*, to which our British Wild Arum or Lords and Ladies, *Arum maculatum*, belongs). Numerous species of *Anthurium* and *Philodendron* climb up or grow on the wire- and peat-covered pillars and up the sides of the house, others are grown in hanging baskets, in pots on the staging or planted out in the beds. Notice the size of the leaves of many of the anthuriums. There are species of *Dracaena* (Lily family, *Liliaceae*) to be seen and some tropical trees, including a large, attractive leguminous tree, *Baikiaea insignis*, from tropical Africa which flowers regularly. It was introduced to Kew in 1894.

🌿 Filmy Fern House F16

The filmy ferns require extremely high humidity and subdued light and to maintain these conditions a north-facing glasshouse within a glasshouse has been constructed with light entering only from above. They are relatively slow-growing and so other ferns have been included. The true filmy ferns, *Hymenophyllaceae*, are small, with fronds only a few or even one cell thick. Here they grow mainly beneath other ferns and on the rock faces, and include *Cardiomanes reniforme* from New Zealand with unusual kidney-shaped fronds. More obvious are species of *Leptopteris* (*Osmundaceae*) with fronds three cells thick but which reach 1 m in length. Both the filmy ferns and *Leptopteris* can usually be recognised by the translucent appearance of the fronds owing to their extreme thinness.

The public area is enhanced with numerous tree-ferns and a collection of smaller ferns including *Osmunda regalis*, the Royal Fern, of interest because the upper portion of the fertile fronds has no leaf blade and consists only of sporangia clustered together.

♣ Princess of Wales Conservatory H15 (UNDER DEVELOPMENT)

This new greenhouse replaces the 26 houses and sections of houses which comprised the 'T' Range and Ferneries complexes. These complexes had developed haphazardly over the years as the need for more space arose, but being made of wood and subjected to a continuously moist, tropical environment, their life-span was as little as thirty years. Recently, the cost of maintaining such houses has increased considerably, thus this new house, in a single structure, represents a major investment which meets all the particular and diverse environmental needs of the tropical collections. From the outset it has been designed to be economic to run and maintain and because it also merges the previously separated displays, more exciting arrangements are possible. These include a mangrove swamp, tropical montane zone, underground viewing aquaria and Namib desert area with a specially heated bed for *Welwitschia mirabilis*. This unique plant, closely related to the conifers, has only two leaves which grow continuously from the base. It is very long-lived and radiocarbon dating suggests that large specimens may be 2000 years old. The traditional displays of ferns, orchids, Giant Waterlily, cacti, succulents, bromeliads, carnivorous plants, begonias etc., are also featured, not forgetting the popular Mohave desert diorama and the *Lithops* (Living Stones) and other desert plants which are sited at the southern entrance.

The conservatory is roughly diamond-shaped in plan and of a low multispan design; it has a strong steel frame (to reduce internal structural elements), and aluminium glazing bars. In many ways, the design reflects the inter-linked, straight-roofed houses of the 'T' Range (demolished in 1983), which originally stood on this site. Another design feature, borrowed from an earlier age, is that much of the interior of this house is below ground level to save heat. Technology has its place however, with a new low-energy humidification system, and a computerised control centre that carefully meters energy usage in the conservatory's ten environments.

Construction of the conservatory began in August 1983 and was completed in 1985. It is hoped that the house will be opened officially in the summer of 1987. Meanwhile, as the planting work is completed, various sections will be open to visitors.

♣ Alpine House J17

The Alpine House, opened in 1981, is sited close to the Jodrell Laboratory and Lecture Theatre (used by Kew's botanical scientists and horticultural students). Shaped like a pyramid, the house is designed to allow the maximum transmission of light and is also exceptionally well ventilated with three flights of ventilators on each face (lifting 55% of the roof surface), and low-level louvres over the moat. The banks of the moat, and

the main internal plantings, are arranged as a sandstone rock garden containing nearly three thousand different montane plants. On a bench at the rear of the house and in the front part of the beds are many pot-grown plants which are changed regularly to provide a beautiful year-round display. Spring is the best season for this house with its miniature iris and cyclamen, primulas and saxifrages, tulips and a myriad other blossoms.

The central bench is refrigerated and fitted with high pressure sodium lamps which provide extra hours of daylight during our short, dull winter. In the larger section, there are plants from the high mountains of the tropics where there is an extreme difference in temperature between each day and night. Thus, with the aid of cooling and lighting, such rarities as the felty-leaved Espeletias from the high Andes, and Lobelias from the African mountains, can now be seen by the visitor.

The smaller section which has additional specialised lighting, is for arctic plants, and conditions here are equivalent to those found in central Greenland. The plants, which include the diminutive Arctic Poppy, *Papaver radicatum*, are transferred in November to a cold store and kept in the dark at 6°C to simulate the conditions in a snowdrift. They are returned to the bench in the following spring.

♨ Palm House H12

The Palm House was designed by Decimus Burton in association with Richard Turner, a civil engineer, and built between 1844–1848. In the 1950's it was found that the structure had deteriorated to the point of making the house unsafe. It was consequently closed for repairs and re-opened in 1959. Since then further deterioration has occurred, and the house was closed in September 1984 to allow for a major restoration, which may take up to 3 years to complete. During the interim period, a representative collection of the Palm House plants can be seen in the Temporary Palm House near the Tea Bar (D14).

♨ Temporary Palm House D13,14

The house contains a collection of tropical plants and includes palms, cycads, screwpines, and many others which are of particular botanical, economic or ornamental interest. The palm collection includes the Coconut, *Cocos nucifera*, the Oil Palm, *Elaeis guineensis*, which yields an edible oil widely used for commercial purposes and the Double Coconut, *Lodoicea maldivica*, from the Seychelles.

The cycads are descendants of an ancient group of plants which flourished over 100 million years ago. They are more closely related to the conifers than to any other group. A remarkable collection of these plants, which today are of very restricted distribution, can be seen here.

The screwpines, *Pandanus*, get their common name from the spiral arrangement of their leaves and from the resemblance of the fruits to those of the pineapple. They are found in the eastern tropics in sea-coast and swampy areas, where their numerous stilt or prop roots help to keep them upright.

Many useful plants can be seen including specimens of banana, cocoa, coffee, rubber, pawpaw, etc; some of these bear a descriptive label. The Giant Bamboo, *Gigantochloa verticillata*, a member of the Grass family, is used for construction work in Java, and grows very rapidly in one season.

Some outstanding ornamental plants include *Brownea*, a tropical tree which bears clusters of beautiful orange flowers, the Jade Vine, *Strongylodon macrobotrys*, with wonderful trusses of blue-green flowers, which is almost extinct in its native Philippines, and many others with brilliant flowers or foliage such as *Bougainvillea*, *Clerodendrum* and *Thunbergia*. Because of their curiously shaped mottled-brown flowers, *Aristolochia* species are sometimes given the name Dutchman's Pipe. Their flowers have a most unpleasant smell and are pollinated by flies, which are attracted by the odour.

✿ Waterlily House (NO. 15) G13

This delightful small greenhouse, which lies immediately north of the Palm House, was built in 1852 by Richard Turner (1798–1881). Intended primarily to house waterlilies, it also contains other beautiful tropical plants including many climbers such as *Allamanda cathartica* and *Clerodendrum thomsonii* which are in flower from May to September in the entrance vestibule.

Most waterlilies belong to the genus *Nymphaea* and these are grown in containers submerged in the pool. The Giant Waterlily, *Victoria amazonica*, which has leaves reaching over 1 m in diameter, is now grown in the Princess of Wales Conservatory. The Sacred Lotus, *Nelumbo nucifera*, native to India, China and Japan, is grown round the island or in one of the side beds. The beautiful rosy-pink flowers are held high above the water on long stalks. In India the lotus is often grown outside temples, where the stems are made into wicks and tapers for use at Hindu ceremonies. Growing on the island is the Papyrus or Paper Sedge, *Cyperus papyrus*; the ancient Egyptians used the stems to make paper, boats, sandals and ropes. A beautiful tropical creeper, *Ipomoea mauritiana*, which has pinkish-mauve flowers, climbs along the rail surrounding the pool.

The plants in the small side pools include rice, *Oryza sativa* and the cocoyam, *Xanthosoma violaceum*. Cotton, *Gossypium* species, can be found growing in the side beds together with jute, *Corchorus olitorius*, black pepper, *Piper nigrum*, pineapple, *Ananas comosus*, and sweet potato, *Ipomoea*

Banana in fruit ▶

batatas. Three varieties of the Rosy Periwinkle, *Catharanthus roseus*, which has been found to contain an important drug useful in the treatment of certain cancers, are also displayed in these beds.

Members of the Cucumber family, *Cucurbitaceae*, are grown up the sides of the house. The Bottle Gourd, *Lagenaria siceraria*, produces fruits which are used to make bottles, bowls, spoons and parts of musical instruments. The young gourds can be bound with strips of cane, etc, to produce the shapes required. The Loofah or Towel Gourd, *Luffa cylindrica*, is well-known for its fibrous fruits used as bath sponges; the fruits are also used in the manufacture of sandals, door and bath mats, and because of their shock-absorbing properties, in steel helmets and armoured cars.

The young fruits of the Bitter Gourd, *Momordica charantia*, and the Snake Gourd, *Trichosanthes cucumerina*, are used as vegetables in some tropical countries. The Wax Gourd, *Benincasa hispida*, native to Java, produces large fruits covered with white wax.

☙ Australian House H6, 7

The Australian House contains a miniscule proportion of the total 20,000 plant species of Australia, but nonetheless conveys the uniqueness of a flora of which some 85 per cent of the species are endemic (i.e. found nowhere else).

The collection includes a number of *Banksia* species. The first banksias were collected as dry specimens by Joseph Banks (later George III's scientific adviser on Kew) in 1770 on Cook's first voyage to Australia and named in his honour. There are also a number of wattles or mimosas (*Acacia* species), and between late December and January the fine tree of *Acacia baileyana* produces a mass of yellow flowers.

Dominating the Australian landscape are the Gum Trees, *Eucalyptus*, which range from small shrubs less than 1 m high to giant trees over 100 m. Natural fire is a common hazard in Australia and some species of *Eucalyptus*, known as mallees, possess a swollen mass of woody tissue (lignotuber) just beneath the ground. This lignotuber contains many dormant buds which grow rapidly after a fire. Other plants such as *Hakea* and the bottle-brushes, *Callistemon*, have very hard woody fruits which remain closed on the plant sometimes for several years, until fire cracks them open so releasing the seeds. Kangaroo Paws, *Anigozanthos*, and the Grass Tree, *Xanthorrhoea australis* which yields a gum used for varnishes, are also to be found in this house.

☙ Temperate House J6,7

The Temperate House, designed by Decimus Burton, was in its day the

largest greenhouse in the world. Constructed over a long period, (Main Block and Octagons 1860–62, Wings 1896–9), it offered a haven of greenery and delicate blooms in winter and a cool oasis in summer beneath tree ferns, palms and tall trees.

In 1972, a survey revealed that the structure and fabric of the building needed major repair and in many cases the complete replacement of the iron members was necessary. The restoration was carried out between 1977–1980 and was so faithful to Burton's designs, that even the intended urns (some missing, some never fitted) were all added to the masonry façade for the first time. The house had a number of design faults, which cumulatively, restricted the plants that could be cultivated. The most serious of these was the loss, caused by the structure, of almost three-quarters of natural light in the Main Block. The restoration has improved this, and removed other problems, to such a degree that the plants are growing better than ever and an even wider range of rare plants is now on display.

The plants are arranged geographically. The South Wing, which naturally catches the most sun, has plants from southern Africa and the Mediterranean region. The South Octagon, which is kept very cool, holds a collection of Cape Heaths or *Erica* species of which there are more than 600 in southern Africa. These beautifully flowered heaths are quite difficult to grow, like their even more problematic neighbours in this house, the members of the family *Proteaceae*.

Within the Main Block there are collections from the Old and New Worlds along with rarities such as *Trochetiopsis* from the South Atlantic island of St. Helena. These trees, distant relatives of the cocoa, occur naturally on St. Helena; *T. erythroxylon* is rare enough (only two plants remain in the wild) but *T. melanoxylon* was thought to be extinct until recently when a single specimen was discovered. Our plant was presented by the University of Cambridge Botanic Garden. In addition to the geographic beds, there is a 'Citrus Walk', where a selection of oranges, lemons and other citrus trees are grown together with tea, *Camellia sinensis*, the Tamarillo or Tree Tomato, *Cyphomandra betacea*, which comes from Peru and Brazil, and the Jojoba, or Goatnut, *Simmondsia chinensis* whose seeds yield a form of liquid wax. Dominant in the centre of the house is a large Chilean Wine Palm (*Jubaea chilensis*) – probably the largest greenhouse plant in existence – which was originally raised from seed at Kew in 1846. This palm withstood the reconstruction protected by a sheath of polythene.

The North Octagon has a fascinating collection of New Zealand's plants including the Poor Knights Island Lily, *Xeronema callistemon*, whilst the North Wing has plants from Asia including an interesting new collection of tender rhododendrons from the mountains of South-East Asia, notably New Guinea.

The restoration of the building, completed in 1980, also provided an exhibition area in the North Octagon basement. The Temperate House was formally re-opened by Her Majesty The Queen on 13th May, 1982.

Lakes, ponds and wildfowl

♨ Palm House Pond H–J12–13

In the time of George III the Palm House Pond covered a larger area, extending westwards and including the site of the Palm House. At that time it was the only lake on the Kew estates. It was partly filled up in 1814, and then given its formal layout by W. A. Nesfield during the construction of the Palm House in 1847.

The fish in the pond include eel, carp, roach and dace. At the edge of the pond are several fine specimens of Swamp Cypress, *Taxodium distichum*, with the woody projections from the roots ('knees') showing clearly. The fountain includes a statue of Hercules and Achelous sculpted by Bosio and cast in 1826 for George IV. It was at Windsor Castle until presented to Kew by HM The Queen in 1963.

♨ The Lake D6 to F8

The Lake is artificial in origin, having been excavated between 1856–61, and enlarged in 1871 to its present size of *c.* 2 hectares (4½ acres). The gravel taken from here was used to make paths and construct mounds on the otherwise flat land of Kew, including the terrace on which the Temperate House stands. In and around the Lake are both introduced and native, water-loving plants.

♨ Waterlily Pond E5

This pond was constructed from a disused gravel pit in 1897. In summer there is a fine collection of waterlily hybrids, best viewed in the mornings when their flowers are fully expanded. The pond is surrounded by water-loving plants, and the notable Swamp cypress, *Taxodium distichum*, growing out of the pond has been previously mentioned (p. 31).

♨ Aquatic Garden J16

The Aquatic Garden is a sunken, formal area near the Grass Garden. It consists of a central tank of waterlilies, flanked on either side by a bed containing a collection of British bog plants. At each corner of the garden, British and other water plants are grown in tubs immersed in small ponds.

These include Bogbean, *Menyanthes trifoliata*, which produces beautiful pinkish flowers in May–June, and the true Bulrush, *Schoenoplectus lacustris*, the stems of which are used for the rush seats of chairs.

The garden is surrounded on three sides by a hedge of *Berberis* × *stenophylla*, which despite vigorous yearly pruning, produces a mass of golden flowers each spring.

♨ Wildfowl

There is a collection of wildfowl at Kew, some of which live on and around the Lake and others by the Palm House Pond. Unfortunately many birds have been lost recently owing to a severe outbreak of botulism. This disease occurs on muddy shallows around ponds and lakes and is most prevalent during hot weather. One of the main contributing factors to the development of this disease is believed to be the bread which is thrown into the water to feed the birds. Notices are therefore displayed asking visitors not to feed them.

Mallards, moorhens, coots and tufted ducks can usually be seen; these birds are not maintained by the Gardens but come and go as they please. Several species of pheasants also breed and roam freely through the grounds.

Index

Page numbers are shown in black type

Map references are shown in blue type

†closed to visitors
*closed in winter

56